THE NUMBER STORY

SMALL BOOK ONE
ENGLISH - SCOTS

Numbers Teach Children Their Number Names

written and illustrated by

MISS ANNA

Early Reader Edition of *The Number Story 1*
Bronze Medal Winner, 2016 Wishing Shelf Book Award

Library of Congress Control Number: 2018902040

Names: Miss Anna, author.
Title: Number story : numbers teach children their number names / Miss Anna.
Description: Portland, OR: Lumpy Publishing, 2018.
Identifiers: ISBN 978-1-949320-08-4| LCCN 2018902040
Summary: The pictures and rhymes present stories which introduce numbers 0-10.
Subjects: LCSH Numeration—English—Scots--Pictorial works--Juvenile literature. | BISAC JUVENILE NONFICTION /
Languages: English—Scots
Classification: LCC QA141.3 .M57 2018 | DDC 513—dc23

Publisher: Lumpy Publishing
Website: www.missannabooks.com
Email: missanna@missannabooks.com

Paperback: ISBN 978-1-949320-08-4
Printed in the U.S.A.   1 3 5 7 9 10 8 6 4 2

Wid ye like to ken
the names of numbers?

It is very easy and a lot of fun!

It's awfy cannie an' loads of fun!

Say-along our little jingle

Sing wae us oor wee story!

starting from Number One!

We will begin fae number wan!

ONE looks like my one finger.

WAN

looks like my wan finger.

ONE!
WAN!

2

TWO   trails a tail.

TWA

trails a tail.

A TAIL! A TAIL!

3

THREE has bumps.

THREE

hae bumps.

BUMPY! BUMPY!

4

FOUR carries a sail.

FOWER

cairies a sail.

A SAIL!
A SAIL!

5

FIVE is a racing track.

FIVE is a racing track.

VROOM
VROOM!

SIX   curves like a snail.

SAX

curvy like a snail.

A SNAIL!   A SNAIL!

7

SEVEN has a sharp angle.

SEIVEN

hae a shairp angle.

BE CAREFUL! IT'S SHARP!

CAW CANNIE! IT'S SHAIRP!

8

EIGHT is rollercoaster rails.

EIGHT

is rollercoaster rails.

YAY!
YIPPEE!

9

NINE   is a bubble on a stick.

NINE

is a bubble on a stick.

A BUBBLE!
A BUBBLE!

10

TEN   is an eye of a whale.

TEN

is yin ee of a whaul.

WINK!
WINK!
HELLO! HULLO!

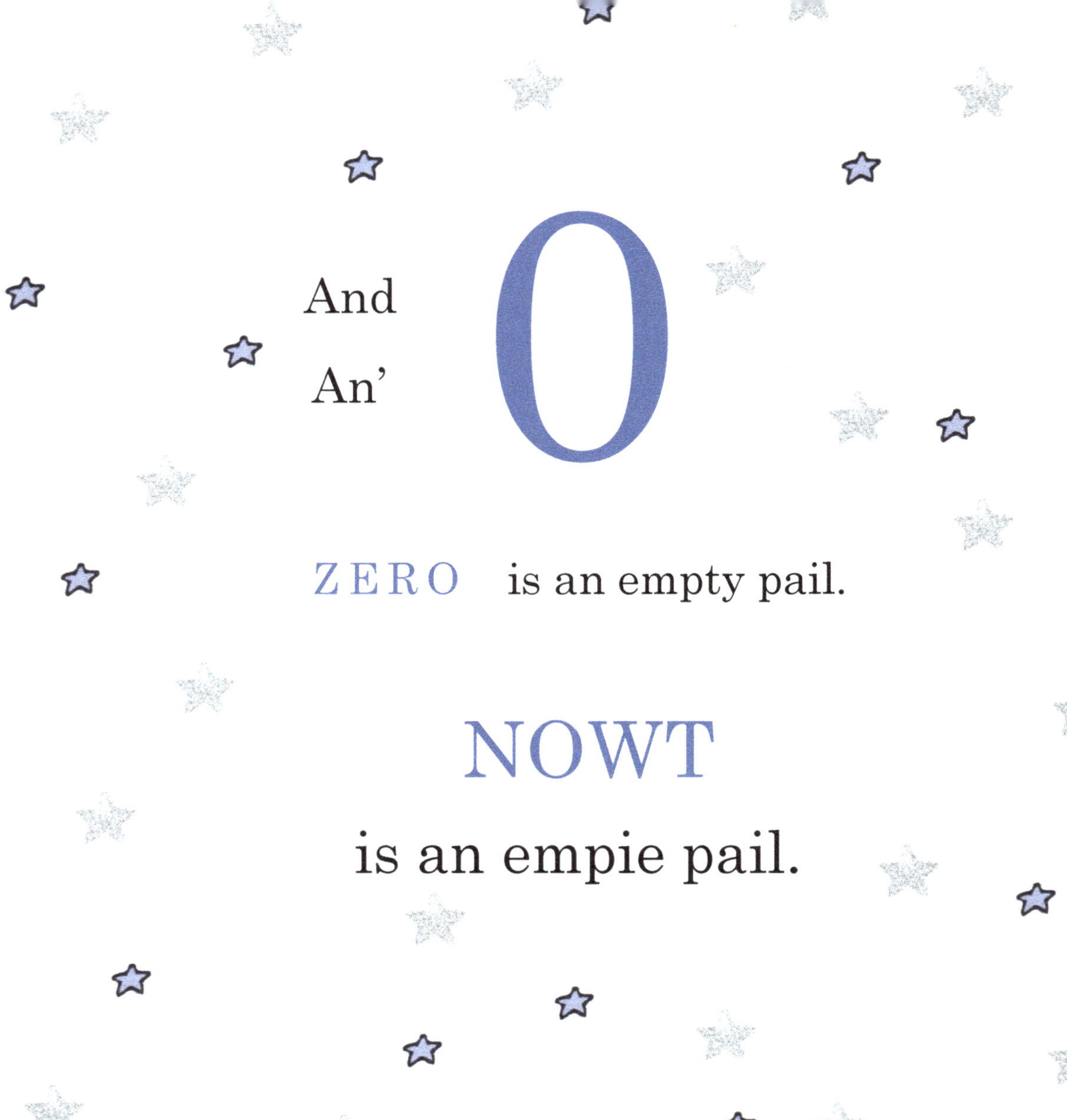

And
An'

O

ZERO   is an empty pail.

NOWT

is an empie pail.

IT'S
EMPTY!
IT'S EMPIE!

Thank you for playing with us today.

We had a lot of fun too!

Thank you for playing wae us the day.

We haed a lot of fun too!

We are your Number friends,
Zero to Ten,
Who will be here for you~
We are your number freends
Nuthin tae Ten.
We will ayeways be here for you!

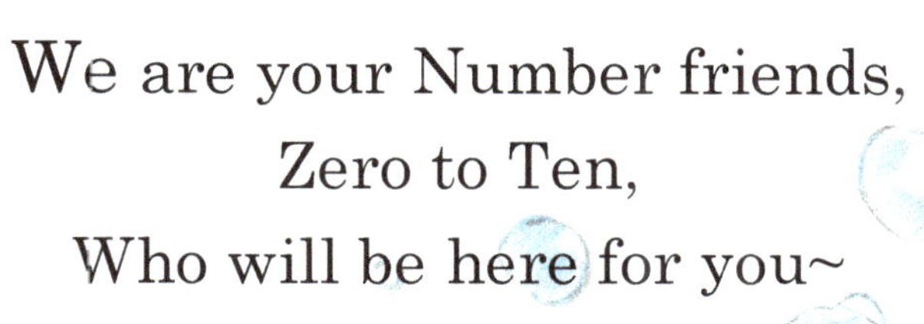

Bye-bye now!
See you again soon!
Cheerio the nou!
Haste ye back!

The Numbers are *SINGING* too!

To sing-a-long, look for Miss Anna Number Story
at your favorite music store like iTUNES.

MP3

Numbers 0-10
IDENTIFYING
& COUNTING

Numbers 11-20
& Ordinals
first, second, third...

Numbers 0-100
& Place Values
ones, tens, hundreds...

About Clocks
& Telling Time
hours, minutes, seconds

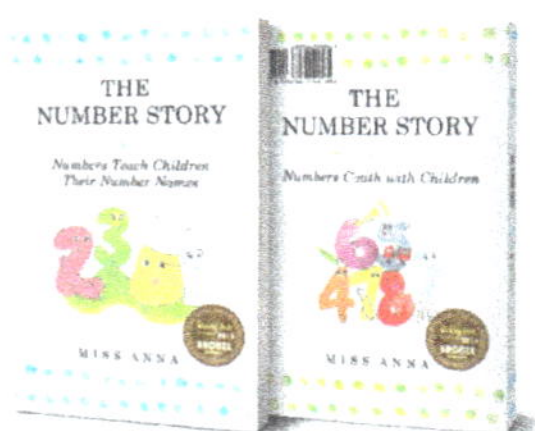

Number Story 1 & 2
isbn: 978-0-996216-48-7

Number Story 3 & 4
isbn: 978-1-945977-01-5

Number Story 5 & 6
isbn: 978-1-945977-06-0

Number Story 7 & 8
isbn: 978-1-949320-40-4

For more Miss Anna books to love,
visit us at

www.missannabooks.com

Numbers are working hard all over the world!
Come Travel the World with Us!

* 9 7 8 1 9 4 9 3 2 0 0 8 4 *